OUR FAVORITE BRANDS

PIXAR

By Martha London

Kaleidoscope
Minneapolis, MN

The Quest for Discovery Never Ends

....

This edition is co-published by agreement between Kaleidoscope and World Book, Inc.

Kaleidoscope Publishing, Inc.
6012 Blue Circle Drive
Minnetonka, MN 55343 U.S.A.

World Book, Inc.
180 North LaSalle St., Suite 900
Chicago IL 60601 U.S.A.

All rights reserved. No part of this book may be reproduced in any form without written permission from the publishers.

Kaleidoscope ISBNs
978-1-64519-019-6 (library bound)
978-1-64494-184-3 (paperback)
978-1-64519-119-3 (ebook)

World Book ISBN
978-0-7166-4319-7 (library bound)

Library of Congress Control Number
2019939232

Text copyright ©2020 by Kaleidoscope Publishing, Inc. All-Star Sports, Bigfoot Books, and associated logos are trademarks and/or registered trademarks of Kaleidoscope Publishing, Inc.

Printed in the United States of America.

FIND ME IF YOU CAN!

Bigfoot lurks within one of the images in this book. It's up to you to find him!

TABLE OF CONTENTS

Chapter 1: A Story for Everyone.. *4*

Chapter 2: A New Way to Make Movies....................... *10*

Chapter 3: To Infinity and Beyond................................... *16*

Chapter 4: Leader of the Pack.. *22*

Beyond the Book... *28*

Research Ninja.. *29*

Further Resources... *30*

Glossary.. *31*

Index... *32*

Photo Credits.. *32*

About the Author... *32*

CHAPTER 1

A Story for Everyone

Anthony and his family walk into the movie theater. Anthony drags his mother by the hand. He wants to find a front-row seat. They sit in big chairs. Anthony holds onto his snack box. It has popcorn, soda, and candy. He saves the candy for last.

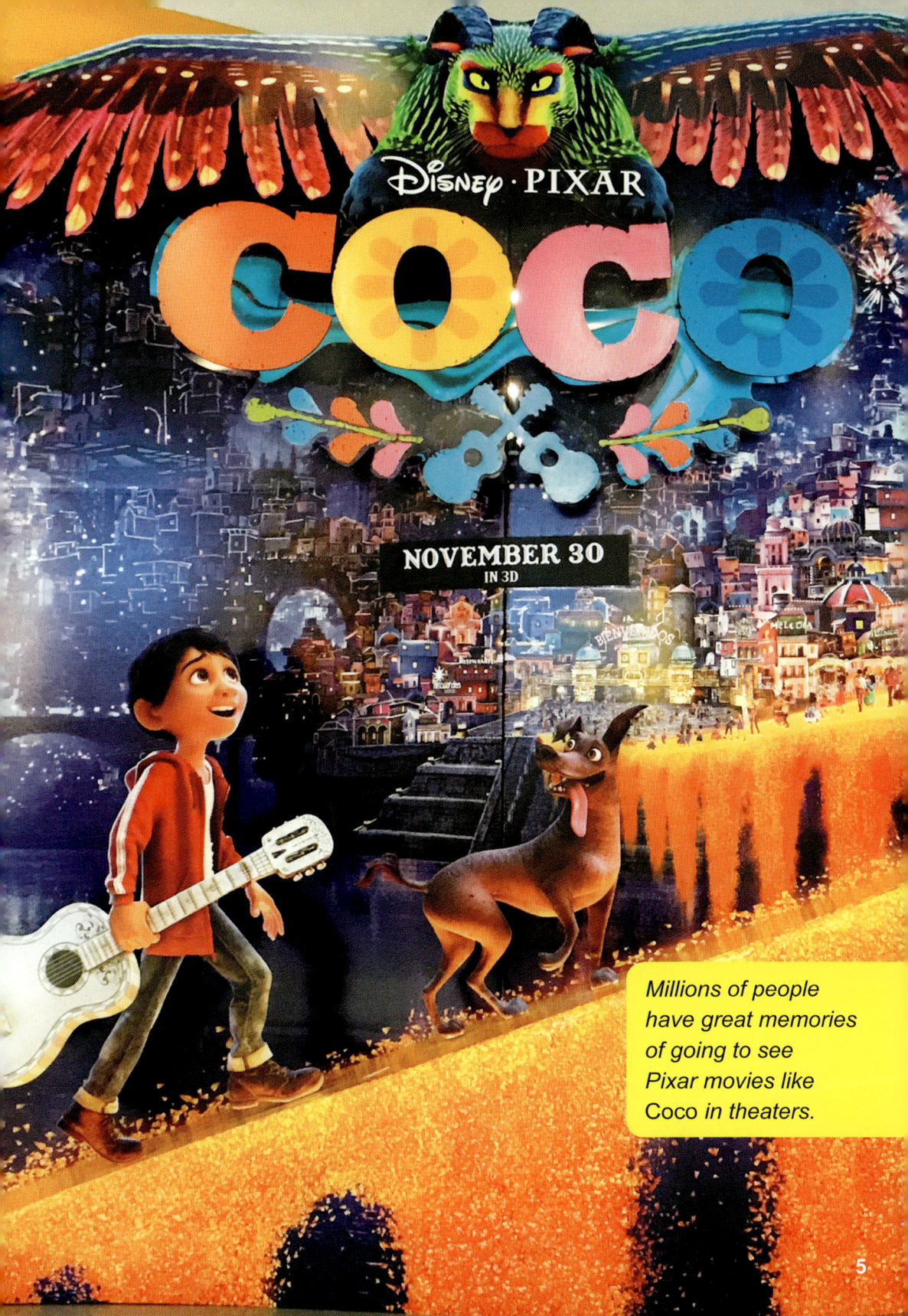

Millions of people have great memories of going to see Pixar movies like Coco in theaters.

Anthony and his family are seeing the new Pixar movie, *Coco*. He's excited because it's about the Day of the Dead. His family celebrates the holiday every year. Anthony hasn't seen a movie about it. The main character is named Miguel. He looks like Anthony. He has tan skin and black hair. Anthony's family is from Mexico. His grandparents still live in Mexico.

The theater is crowded. There are families like Anthony's. There are adults here by themselves. Anthony sees many other Mexican American people. Maybe they are as excited as he is.

Soon the lights dim. Anthony squishes down into his chair. His mom is on one side. His dad is on the other.

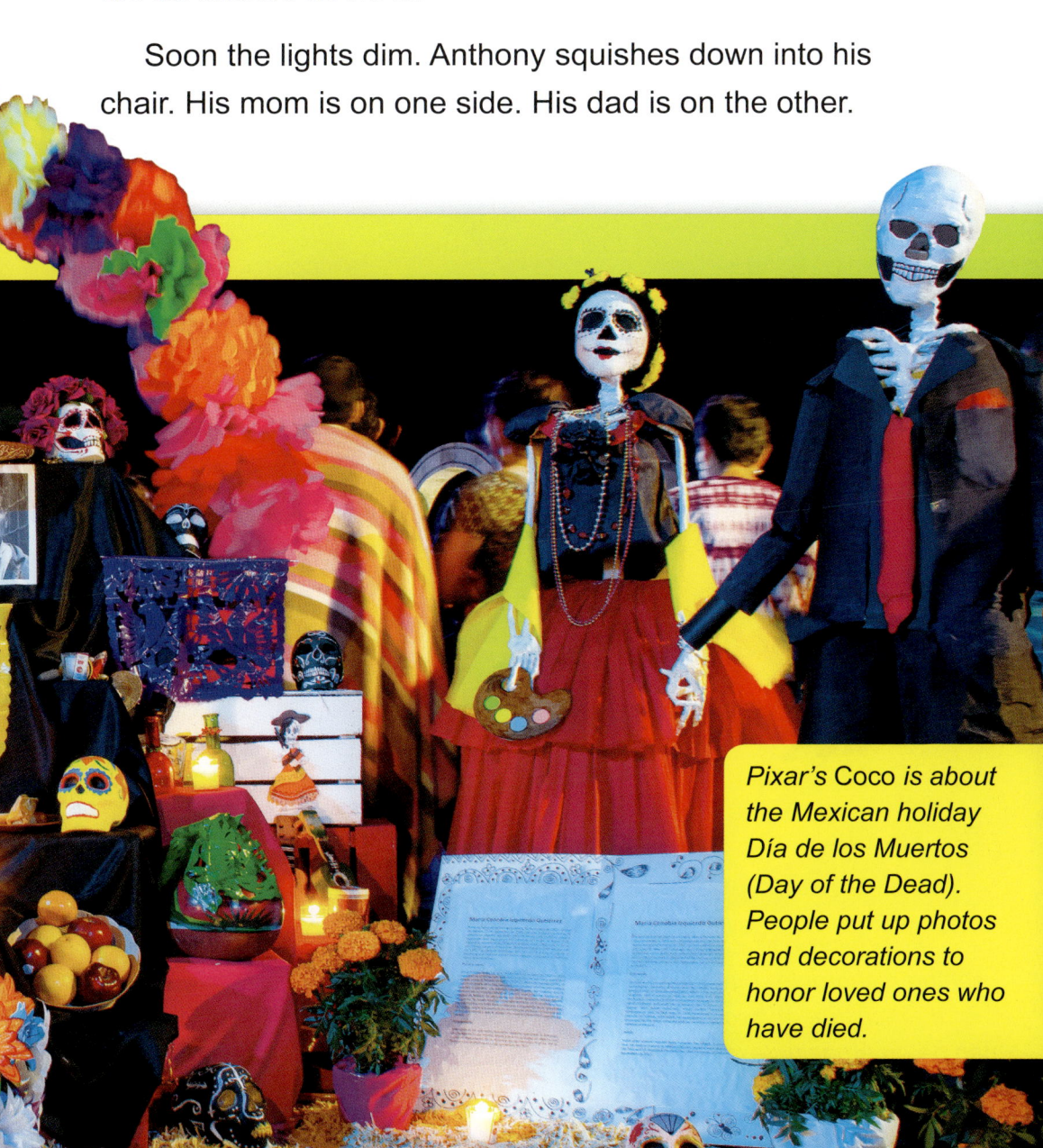

Pixar's Coco *is about the Mexican holiday Día de los Muertos (Day of the Dead). People put up photos and decorations to honor loved ones who have died.*

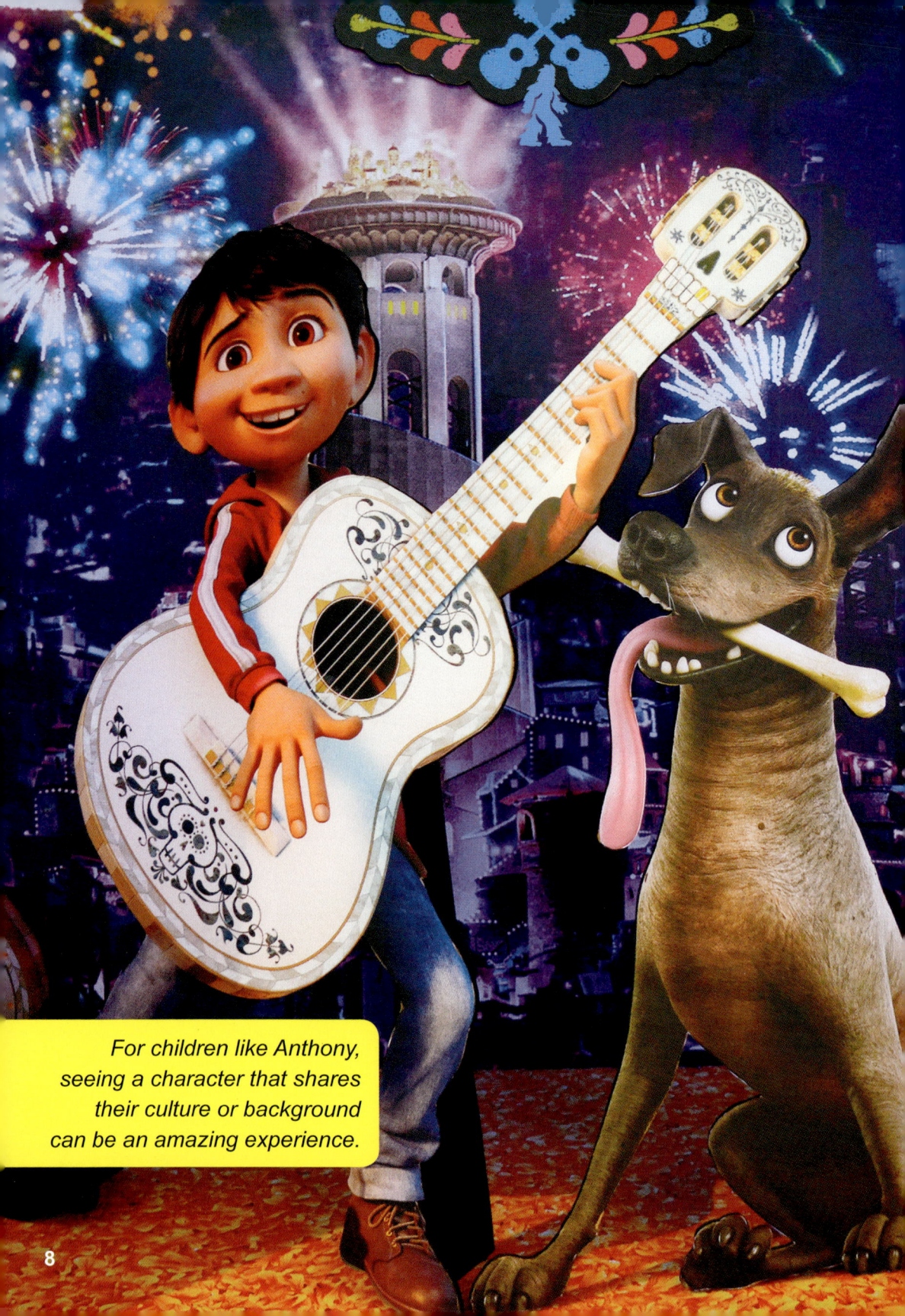

For children like Anthony, seeing a character that shares their culture or background can be an amazing experience.

The movie is full of bright colors. Everything is **vibrant**. Miguel plays the guitar. Anthony loves the music. It sounds like the music his dad plays on the radio. Anthony's house looks different from Miguel's. But the table of photos is familiar. His family puts one up every year. The photos are to remember people who died.

Anthony hardly blinks during the movie. When it ends, the lights turn on. He looks up at his mom and dad. They smile at him. He smiles back. He says he's going to learn to play guitar. Anthony wants to be just like Miguel.

FUN FACT
Dante, the dog in *Coco*, is a Xolo dog, the national dog of Mexico.

CHAPTER 2

A New Way to Make Movies

In 1986, Pixar animators had an idea. Pixar was a new **animation** company. The animators wanted to show how computers could animate shadows. They would make a short film. It would be called *Luxo Jr.* They wanted to show it off. They planned to play it at a **graphics** conference.

This event was called SIGGRAPH. The animators hoped the film would impress other companies.

SIGGRAPH was coming up soon. The animators had to work day and night. One brought a sleeping bag to work. Each night, he slept under his desk. When he woke up, he got back to work.

Pixar Animation Studios started in 1986. In 2000, it moved to its new building in Emeryville, California.

Pixar shows its pride in Luxo Jr. *with a statue outside its studios.*

The film was about two lamps. One of them was smaller. The lamps played with a ball. The little lamp jumped on the ball until it popped.

Luxo Jr. was finished in time. It played at SIGGRAPH. The crowd loved it. Before the film had even ended, people stood. They clapped and clapped. The audience liked the animation. They loved the lamps' story, too. Pixar's short film was a success.

Pixar started making its first full-length movie in 1991. The team had to choose characters. Computers made everything look plastic. Toys would be the perfect characters. *Toy Story* began its journey.

The designers weren't sure how viewers would react. A full-length computer-animated movie hadn't been made before. But the team didn't worry too much. They just needed a good story. Then the kind of animation wouldn't matter.

MAKING A MOVIE

Computer-animated movies can take years to finish. They require big teams. Pixar has 600 employees. Some characters take a long time to animate. One example is Sully from *Monsters, Inc.* He is covered in thick fur. Each **frame** that he was in took eleven hours to create.

At first, the story wasn't good. The writers hunched over their computers. They worked long hours. But they didn't give up. They fixed the **script**. Artists started animating.

Pixar finished *Toy Story* in 1995. Everyone on the team held their breath. It was time to see what the public thought.

The movie was a hit! Pixar got an award for its **innovation**. The company kept making movies. It continued to amaze people. Pixar turned 30 years old in 2016.

Toy Story *was such a success that Pixar developed sequels. Beloved characters attended the premiere of* Toy Story 3.

CHAPTER 3

To Infinity and Beyond

Toy Story became a film **franchise**. Other Pixar movies became franchises, too. The movies turned into books, toys, and games.

Maria and Carter go to the store with their mom. They ask if they can see the new *Toy Story* toys. Their mom nods. She follows behind Maria and Carter. They walk fast. Their mom says they shouldn't run.

In the aisle, the clear plastic sparkles. Everything is shiny and clean. Carter heads for Buzz Lightyear. He can press buttons on Buzz's suit. The buttons make Buzz speak. He sounds just like the movie. Maria prefers Woody. She wants to dress up as a cowgirl for Halloween.

Toys based on the characters from Toy Story *are popular around the world.*

FROM SCREEN TO SHELF

Pixar makes a lot of money from ticket sales. But some movies do even better outside the theater. They make more money from merchandise sales than at the box office. *Toy Story* and *Cars* both did this. The *Cars* franchise has made $10 billion from product sales. There are *Cars* toys, clothes, and more.

FUN FACT

The "voices" of both WALL-E in *WALL-E* and R2-D2 in *Star Wars* were created by the same sound designer.

Natalie and Tina are best friends. They watch Pixar movies together. Their favorite is *Finding Dory*. They have the *Finding Dory* version of UNO. Natalie and Tina play a lot of UNO at sleepovers. The cards are worn. Some of the edges are torn. But Natalie and Tina don't care. They love seeing their favorite characters.

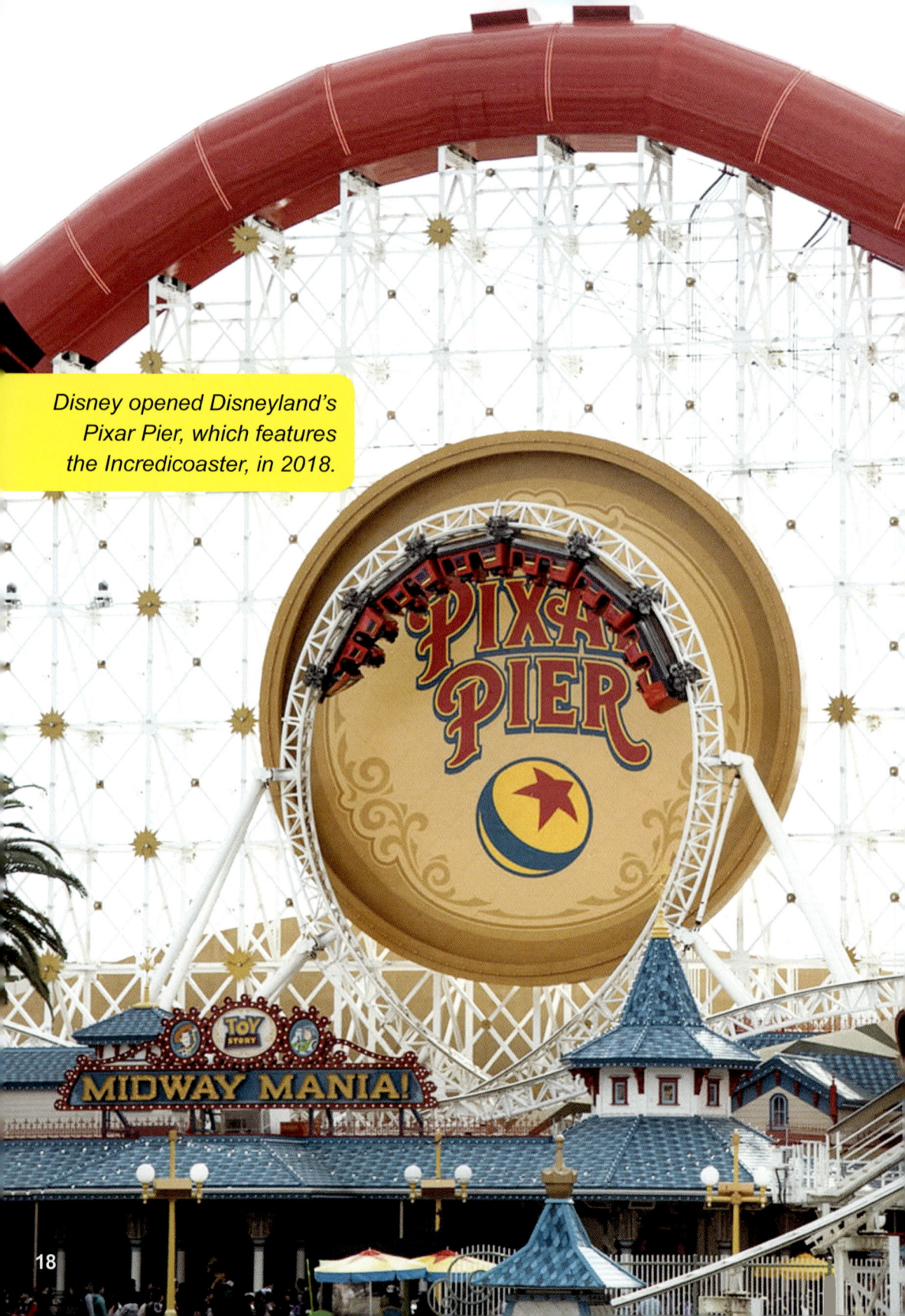

Disney opened Disneyland's Pixar Pier, which features the Incredicoaster, in 2018.

Raj can't wait for his birthday. His family is going to Disneyland in California. They're going to visit the new Pixar Pier. Raj is excited to ride the Incredicoaster. It has characters from *The Incredibles* along the ride. He thinks the roller coaster will be as fast as Dash.

FUN FACT
The original title of *The Incredibles* was *The Invincibles*.

PIXAR TIMELINE

1986
Luxo Jr. premieres at SIGGRAPH.

1996
Disney and Pixar announce an agreement to produce five movies together over ten years.

1999
Toy Story 2

2003
Finding Nemo

2006
Disney buys Pixar.

1986   1996   1998   2000   2002   2004   2006

1995
Toy Story

2001
Monsters, Inc.

1998
A Bug's Life

2004
The Incredibles

2006
Cars

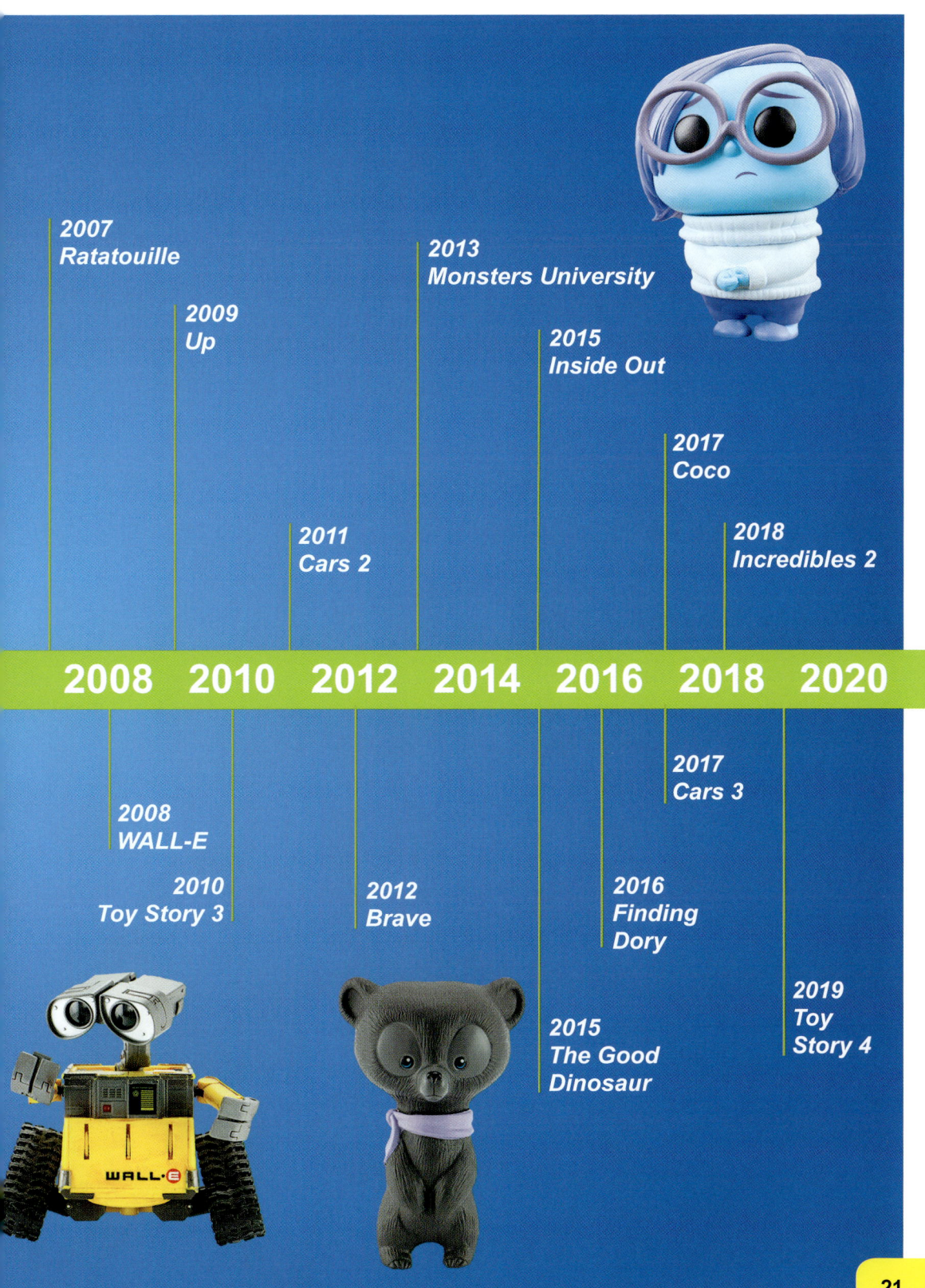

CHAPTER 4

Leader of the Pack

Toy Story was the first entirely computer-animated film. No one had done that before. People weren't even sure it was possible. But Pixar changed that. It opened the door for other filmmakers. Other companies made their own computer-animated films.

Pixar also changed the way stories were told. Pixar knew that its stories needed to last longer than the technology. Technology changes quickly. Creators don't want their movies to lose their magic. Pixar realized the best way to prevent that was to write great stories.

Models of Monsters University *characters Mike and Sully were featured in a museum exhibit called "The Science Behind Pixar."*

PIXAR AT THE BOX OFFICE

in Millions of Dollars

Film	Box Office
Ratatouille	$470.7
The Incredibles	$541
Monsters University	$544.2
Up	$560
Coco	$632.1
Inside Out	$682.6
Finding Dory	$828.5
Finding Nemo	$846.3
Toy Story 3	$866.9
Incredibles 2	$964.8

These great stories help people relate to characters. Sometimes, they help people feel understood. When Anthony saw *Coco,* he couldn't believe it. He'd never seen a movie about a child that looked like him.

David saw *Inside Out* in the theater with his parents. He had been feeling down for a while. One of his best friends moved away. David saw himself in Riley. He didn't know how to talk about his feelings. But *Inside Out* helped him. After the movie, he talked to his mom.

The characters from Pixar's Inside Out *can help viewers understand and talk about emotions like sadness.*

An actress attended the release of Brave dressed as Princess Merida with her hair styled to look like the detailed animated curls from the movie.

They sat on the couch together. She explained it was okay to feel sad. David and his mom called David's friend. They talked on the phone. Talking to his friend and his mom helped David feel better. *Inside Out* gave him a new way to think about these things.

Pixar pushed the limits of technology. Pixar's team kept improving. Soon, movies didn't have to be about plastic toys. They could make Merida's coiled hair in *Brave*. Backgrounds could look realistic. Characters could live underwater or in space. A house could be lifted by balloons. Thanks to Pixar, animated stories can feel real.

FUN FACT
Pixar animated more than 10,000 balloons to carry Carl's house in *Up*.

BEYOND THE BOOK

After reading the book, it's time to think about what you learned. Try the following exercises to jumpstart your ideas.

THINK

DIFFERENT SOURCES. Consider the different types of sources you could use to research Pixar. How might each type of source be useful in its own way?

CREATE

SHARPEN YOUR RESEARCH SKILLS. *Coco* is about the Mexican holiday Day of the Dead. Where could you find more information about the Day of the Dead? Write a paragraph about your next steps for research.

SHARE

WHAT'S YOUR OPINION? This book says that Pixar changed the way movies were made. Do you agree or disagree? Find evidence in the text that supports your position. Share your position and evidence with a friend. Does your friend find your argument convincing?

GROW

REAL-LIFE RESEARCH. What kinds of real-world places could you visit to find more information about computer animation? What other topics could you explore at these places?

RESEARCH NINJA

Visit **www.ninjaresearcher.com/0196** to learn how to take your research skills and book report writing to the next level!

RESEARCH

DIGITAL LITERACY TOOLS

SEARCH LIKE A PRO
Learn about how to use search engines to find useful websites.

FACT OR FAKE?
Discover how you can tell a trusted website from an untrustworthy resource.

TEXT DETECTIVE
Explore how to zero in on the information you need most.

SHOW YOUR WORK
Research responsibly—learn how to cite sources.

WRITE

GET TO THE POINT
Learn how to express your main ideas.

PLAN OF ATTACK
Learn prewriting exercises and create an outline.

DOWNLOADABLE REPORT FORMS

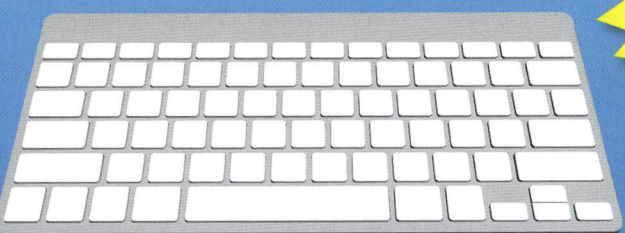

Further Resources

BOOKS

Bishop-Stephens, Will. *How to Create Animation in 10 Easy Lessons*. QED Publishing, 2016.

Cooper Hewitt, Smithsonian Design Museum. *Designing with Pixar: 45 Activities to Create Your Own Characters, Worlds, and Stories*. Chronicle Books, 2016.

Lyons, Heather. *Kids Get Coding: Programming Games and Animation*. Lerner, 2018.

WEBSITES

Factsurfer.com gives you a safe, fun way to find more information.

1. Go to www.factsurfer.com.
2. Enter "Pixar" into the search box and click 🔍.
3. Select your book cover to see a list of related websites.

Glossary

animation: Animation is drawing pictures by hand or on a computer and showing them back-to-back so that they seem to move. The team at Pixar wanted to show the effects they could create with computer animation.

frame: A frame is a single image from a movie. Each frame in *Monsters, Inc.* that had Sully in it took many hours to draw.

franchise: A franchise is a collection of movies, books, and/or television shows about the same characters or fictional universe. Kids can buy toys based on the characters from the *Toy Story* franchise.

graphics: Graphics are images made on a computer. Audiences at the movie *Toy Story* were impressed by how realistic the graphics were.

innovation: Innovation is doing things in a new way. Pixar received awards for its innovation in computer animation.

merchandise: Merchandise is a collection of products that people can purchase. The movie *Cars* made more money from merchandise sales than it did from ticket sales.

script: A script is the written lines that actors say in a movie. The writers at Pixar had to make many changes to the *Toy Story* script before they were ready to make the movie.

vibrant: A color that is vibrant is bright and vivid. The movie *Coco* is animated with beautiful, vibrant colors.

Index

animation, 10–11, 13–14, 22, 27

Brave, 21, 27

Cars, 17, 20–21
Coco, 4, 6–7, 9, 21, 23, 24

Day of the Dead, 6, 9
Disney, 19, 20

Finding Dory, 17, 21, 23

Incredibles, The, 19, 20–21, 23
innovation, 14, 22, 27
Inside Out, 21, 23, 24, 27

Luxo Jr., 10, 13, 20

merchandise, 16–17
Monsters, Inc., 14, 20
Monsters University, 21, 23

Pixar Pier, 19

technology, 10–11, 22, 27
Toy Story, 13–14, 16, 17, 20–21, 22, 23

Up, 21, 23, 27

WALL-E, 17, 21

PHOTO CREDITS

The images in this book are reproduced through the courtesy of: cjp/iStockphoto, front cover (Buzz Lightyear); Music4mix/Shutterstock Images, front cover (car); LungLee/Shutterstock Images, front cover (monster); Nicescene/Shutterstock Images, 3, 17 (top), 20 (top), 21 (bottom left), 21 (bottom right); melissamn/Shutterstock Images, 4, 21 (top), 24–25; Sarunyu L/Shutterstock Images, 4–5, 8–9, 19; Carlos Ivan Palacios/Shutterstock Images, 6–7; andresr/iStockphoto, 9; Beto Chagas/Shutterstock Images, 10–11; David Paul Morris/Bloomberg/Getty Images, 12–13; Paul Smith/Featureflash Photo Agency/Shutterstock Images, 14; Dominic Lipinski/Press Association/PA Wire URN:9191598/AP Images, 15; Mercury Green/Shutterstock Images, 16; Mirco Vacca/Shutterstock Images, 17 (bottom); Ed Ruvalcaba/IOS/AP Images, 18–19; Music4mix/Shutterstock Images, 20 (bottom), 30; Joe Hendrickson/Shutterstock Images, 22; kuzina/Shutterstock Images, 23 (film); Craig Russell/Shutterstock Images, 24; Todd Williamson/Invision/AP Images, 26–27.

ABOUT THE AUTHOR

Martha London is a writer from Minnesota. She grew up watching all of Pixar's movies. Martha lives in the Twin Cities.